John Eiesland

On a Certain Class of Functions with Line-Singularities

John Eiesland

On a Certain Class of Functions with Line-Singularities

ISBN/EAN: 9783337179021

Printed in Europe, USA, Canada, Australia, Japan

Cover: Foto ©Thomas Meinert / pixelio.de

More available books at **www.hansebooks.com**

On a Certain Class of Functions with Line-Singularities

DISSERTATION PRESENTED TO THE BOARD OF UNIVERSITY STUDIES OF THE JOHNS HOPKINS UNIVERSITY FOR THE DEGREE OF DOCTOR OF PHILOSOPHY

BY

JOHN EIESLAND

BALTIMORE, 1898

EASTON, PA.:
THE CHEMICAL PUBLISHING COMPANY.
1899.

INTRODUCTION.

The object of this paper is to investigate certain properties of functions with line-singularities. In the first part a class of functions of a real variable has been considered which are closely connected with these transcendentals, and in the case of which MacLaurin's development does not hold, in spite of the fact that all the derivatives of the function are finite as well as the function itself. Such functions were considered by Pringsheim in an article "Zur Theorie der Taylorschen Reihe," *Math. Ann.*, Vol. 42. Attention has been called to certain functions of this kind which all give the same development in a MacLaurin's series, the identity of the different series being due to the periodicity of the coefficients. The correspondence between these functions and the functions represented by their expansion in series has been shown, and also how Poincaré's method of solving an infinite system of equations with an infinite number of unknowns may be applied in expanding such functions. The above-mentioned correspondence was pointed out by Borel in his thesis, "Sur Quelques Points de la Théorie des Functions," Paris, 1894.

In the second part functions of the type

$$f(z) = \sum_{\nu=0}^{\nu=\infty}{}' \frac{c_\nu}{a_\nu - z}$$

have been considered, a_ν being an ensemble forming at most essentially singular lines. Different methods of developing in series have been discussed, development in areas bounded by ellipses and other curves, these being essentially singular lines. It is also shown how Borel's theorem may be applied to such functions, and that they may be continued across the singular lines without loss of continuity, provided the variable is restricted to move in certain directions.

The third part has been devoted to the study of functions possessing the property of remaining finite and continuous as well as all its derivatives on the singular line. It was Prof. Mittag-Leffler who first called the attention of mathematicians to this class of functions in a note in *Acta Math.*, Vol. XV, where he gives an example of such a function constructed by his pupil Fredholm. Pringsheim, in the above-mentioned paper, discusses functions possessing a similar property, and Borel, in his thesis, gives another method of generating these functions. A third method was discovered during the course of my work, in fact, I have shown that functions of the form

$$\Psi(z) = \prod_{\nu=1}^{\nu=\infty} \frac{z-a_\nu}{z-b_\nu} e^{\Phi_\nu(z)},$$

$$\text{where } \Phi_\nu(z) = \frac{(a_\nu - b_\nu)e^{\nu i}}{z-(1+b_\nu)e^{\nu i}} + \frac{1}{2}\left[\frac{(a_\nu - b_\nu)e^{\nu i}}{z-(1+b_\nu)e^{\nu i}}\right]^2 + \cdots$$

$$\cdots + \frac{1}{1-\nu}\left[\frac{(a_\nu - b_\nu)e^{\nu i}}{z-(1+b_\nu)e^{\nu i}}\right]^{\nu-1},$$

can by proper choice of ensembles a_ν and b_ν be made to possess the above-mentioned property. Another simpler type included in the more general one given above, *viz.*:

$$F(z) = \prod_{\nu=1}^{\nu=\infty} \frac{z-a_\nu}{z-b_\nu}$$

has been shown to be closely related to the form given by Pringsheim,

$$f(z) = \sum \frac{c_\nu}{z-b_\nu}.$$

I am indebted to Prof. A. Chessin for valuable suggestions and criticisms offered during the final revision of this paper.

I.

We shall call a function holomorphic in a region, if at all points within it the function is finite and continuous and possesses a finite and continuous derivative. At any point within such a region the function may be represented by a convergent powerseries $P(x - x_0)$, having a radius of convergence greater than zero. Suppose now that x_0 moves continuously from a position x_0 to X. If the path does not pass through a critical point, there will to any position of x_0, say X′, correspond a development of the function in a powerseries $P(x - X')$ with a radius of convergence greater than zero. From this series we may, by the so-called analytical continuation, derive a new series representing the function at a point taken within the circle of convergence of $P(x - X')$, and so on. The particular expression for the function at any point in the path is called by Weierstrass *an element* of the function, and the totality of these elements define a monogenic and analytic function in that part of the plane which is covered by the circles of convergence.

The great majority of analytic functions exist in general throughout the whole plane, except at certain points, the singular points. These may be branch-points, polar, logarithmic, or essential singularities.

The concept of an analytic function necessarily involves the property of possessing finite and determinate derivatives without which no development is possible. That finite and continuous functions do exist which have no derivatives, and thus cannot be represented as an analytic function, Weierstrass has proved beyond any doubt.[1] In fact, it is well-known that the function represented by the series

$$g(z) = \sum b^n z^{a^n},$$

where b is a positive quantity less than unity and a a positive odd integer, converges uniformly and unconditionally for all points within and on the circle $|z| = 1$ and diverges for all points outside; this function, moreover, does not possess any derivatives on the circle of convergence, if $ab > 1 + \frac{3}{2}\pi$, although it is holomorphic for all points within. If we put $z = e^{i\theta}$ and take the real part of the series, we get the function

$$f(\theta) = \sum_{0}^{\infty}{}_n\, b^n \cos(a^n\theta)$$

[1] See Weierstrass, Abh. aus der Funktionenlehre, p. 91.

which is a convergent series and represents a finite and continuous function of a real variable. Weierstrass has shown that this function for no value of θ possesses a determinate differential coefficient, provided $ab > 1 + \frac{3}{2}\pi$. By means of the function

$$\phi(x) = E(x) + \sqrt{x - E(x)},$$

where $E(x)$ stands for the greatest integer in x, Schwarz[1] has constructed a function which is finite and continuous and possessing nowhere a derivative. This function is given by the series

$$f(x) = \sum_{0}^{\infty} {}_n \frac{\phi(2^n x)}{2^n . 2^n}.$$

We may suspect that an indefinite number of such functions exist, which in fact is the case, as has been shown by Lerch, who has constructed a powerseries including Weierstrass' series as a special case (Crelle, Vol. CIII).

The property of possessing finite and determinate derivatives is however not sufficient to make sure that the function is analytic. In fact, given the analytic function $f(x)$ on one side, and on the other the Taylor series

$$\sum \frac{f_\nu(x_0)}{\nu!}(x - x_0)^\nu,$$

the question of convergence of this series naturally presents itself, and even if the Taylor series is convergent, we are *a priori* not at all certain that the development so formed represents the function $f(x)$ itself. Cauchy as early as 1823 advanced this idea in opposition to Lagrange, who claimed that the finiteness and continuity of $f^\nu(x)$ was enough to establish the convergence and validity of the expansion given above. The example used by Cauchy to illustrate this was, however, not a happy one, and some mathematicians[2] consider it insufficient to prove the statement he made concerning the validity of

---

[1] Schwarz, Gesammelte Abhandlungen, II, pp. 279–274.

[2] See Pringsheim, "Zur Theorie der Taylorschen Reihe," *Math. Annalen*, B. 42, pp. 155 and 161.

an expansion. The example given by Pringsheim in the above-mentioned paper illustrates clearly what we have said. In fact, the function

$$f(x)=\sum\frac{(-1)^\nu}{\nu!}\frac{1}{a^\nu x+1},\ a>1 \text{ and real,}$$

when expanded, gives rise to the series

$$\phi(x)=e^{-1}-e^{-a}x+e^{-a^2}x^2-\ldots\pm e^{-a^n}x^n\mp\ldots,$$

which is convergent throughout the finite portion of the plane. That $\phi(x)$ does not represent $f(x)$ is evident from the fact that $f(x)$ in the neighborhood of $x=0$ has an infinite number of poles, $x=0$ being a point-limit of poles, while $\phi(x)$ is holomorphic throughout the whole plane.

We shall now proceed to give a few examples which enable us to put directly into evidence the non-identity of the expansion with the function expanded without resorting to function—theoretical considerations. Pringsheim's attempt in this direction does not seem to be a complete success, on account of the choice of the function necessitating a laborious method of approximation. Consider the series

$$f(x)=\sum_0^\infty\frac{(-1)^\nu\pi^{2\nu+1}}{(2\nu+1)!}\frac{1}{1+a^{2\nu+1}x},$$

where a is a positive integer greater than one. The series is uniformly convergent for all values of x with the exception of the points $-\frac{1}{a}$, $+\frac{1}{a^3}$, ..., situated on the negative half of the x-axis, zero being a point-limit as before. We find

$$f^n(x)=(-1)^n n!\sum_0^\infty{}_\nu\frac{(-1)^\nu\pi^{2\nu+1}}{(2\nu+1)!}\frac{a^{n(2\nu+1)}}{(1+a^{2\nu+1}x)^{n+1}},$$

$$f(0)=\sum_0^\infty{}_\nu\frac{(-1)^\nu\pi^{2\nu+1}}{(2\nu+1)!}=\sin\pi,$$

...

...

$$f^n(0) = (-1)^n n! \sum \frac{(-1)^\nu \pi^{2\nu+1}}{(2\nu+1)!} a^{n(2\nu+1)} = (-1)^n n! \sin a^n \pi,$$

and we arrive at the following expansion:

$$\phi(x) = \sin \pi - \sin a\pi . x + \sin a^2\pi . x^2 - \ldots .$$

Suppose now a = a positive integer greater than unity, each term in this series vanishes. We have thus a function with the same property as Cauchy's well-known $f(x) = e^{-\frac{1}{x^2}}$, but free from all the objections that may be raised against this example. It is also much simpler than the one given by Pringsheim[1]. That $\phi(x)$ does not represent $f(x)$ is obvious from the fact that $f(x)$ is zero for $x = 0$, and differs from zero for all other values of x in the neighborhood of the origin.[2] If instead of π we write $\frac{\pi}{2}$, $f(x)$ becomes

$$f(x) = \sum_\nu \frac{(-1)_\nu \left(\frac{\pi}{2}\right)^{2\nu+1}}{(2\nu+1)!} \frac{1}{1 + a^{2\nu+1} x},$$

and the corresponding expansion is

$$\phi(x) = \sin \frac{\pi}{2} - \sin \frac{a\pi}{2} . x + \sin \frac{a^2\pi}{2} . x^2 - \ldots ,$$

a being subject to the same condition as before. If a is an even integer, we get

$$\phi_1(x) = 1,$$

and when a is an odd integer of the form $4n + 1$, we have

$$\phi_2(x) = 1 - x + x^2 - x^3 + \ldots ,$$

and for a = an odd integer of the form $4n - 1$

$$\phi_3(x) = 1 + x + x^2 + \ldots .$$

---

[1] Pringsheim's example is

$$f(x) = \sum_\nu (-1)^\nu \left\{ \frac{1}{\nu!} \frac{1}{1 + a^{2\nu} x} - \left(\frac{1}{e}\right)^{a^{2\nu} x^{2\nu}} \right\}.$$

[2] The non-identity of $f(x)$ and $\phi(x)$ is also evident, if we apply the well-known theorem that a function cannot be identically zero in any finite part of the plane, or at all points on a line of finite length, without being identically zero throughout the whole plane.

Neither of these expansions represents the respective functions. In fact, $\phi_1(x) = 1 =$ const., while $f(x, 2n)$ is not a constant for all values of x. In the same way, $\phi_2(x) = \frac{1}{1+x}$ and $\phi_3(x) = \frac{1}{1-x}$.[1] The first case illustrates the singular fact that functions do exist, all whose derivatives vanish at a given point, without the functions being constant. In the preceding example the constant happened to be zero. We might easily multiply examples of this kind; thus the function

$$f(x) = \sum_{0}^{\infty}{}_{\nu} \frac{(-1)^{2\nu}}{2\nu!} \left(\frac{\pi}{2}\right)^{2\nu} \frac{1}{1+a^{2\nu}x}$$

gives rise to the expansion

$$\phi(x) = \cos\frac{\pi}{2} - \cos\frac{a\,\pi}{2}\,.\,x + \cos\frac{a^2\pi}{2}\,.\,x^2 - \ldots$$

and making different hypotheses concerning the constant a we arrive at results similar to the above. Again, we may distribute the poles on the positive half of the x-axis and subtract the resulting function from the one already considered; the new function will present the same features as the above.. We may also distribute the poles on the imaginary y-axis and thus derive functions similar to the above. Finally we notice this important fact, to the holomorphic function $f(x)$ corresponds some function $\phi(x)$, but the correspondence is not a one to one, in fact, to a function $\phi(x)$ corresponds an infinite number of functions $f(x)$, while to $f(x)$ corresponds only one holomorphic function. This is entirely due to the periodicity of the coefficients in the expansion. These examples also illustrate the fact that a sum of Taylor's series whose radii of convergence tend to zero may have a finite and even infinite radius of convergence.

---

[1] The non-identity of these expansions may be proved more rigorously thus: take the first case. If we can prove the non-identity of $\phi_1(x)$ and $f(x, 2n)$ when x takes any positive value $> \frac{1}{2n}$ it is evident that the non-identity is proved for all values of x. We have $f(x, 2n) < \frac{\pi}{2}\frac{1}{1+2nx} < 1$, $\left(x \geqq \frac{1}{2n}\right)$, while $\phi_1(x) = 1$. The value $x = 0$ is of course excluded.

We mentioned above that there exists a correspondence between certain functions of the type $f(x)=\sum\frac{c_\nu}{1+a^\nu x}$, having different sets of poles, and their corresponding expansion due to the periodicity of the coefficients of the expansion. We shall now show that a correspondence of a somewhat similar nature exists in the case of functions of the same type, all having the same poles, but different sets of coefficients c_ν.[1]

Let

$$f(x)=\sum\frac{c_\nu}{1+a_\nu x}$$

where a_ν is an ensemble, such that $\lim_{\nu=\infty} a_\nu=\infty$. In order that this function shall be of the kind we have been discussing, we must have

$$(1)\quad \sum_0^\infty{}^\nu c_\nu = A_0,\ \sum_0^\infty{}^\nu c_\nu a_\nu = A_1,\ \ldots,\ \sum_0^\infty{}^\nu c_\nu a_\nu^n = A_n,$$

where the A's are fixed quantities, such that the series

$$\phi(x) = A_0 - A_1 x + A_2 x^2 - \ldots$$

shall be absolutely convergent. If then solutions c_ν of the system (1) can be found, then will $f(x)$ have the required property. But, as Poincaré has shown,[2] we can always solve such a system, and since these solutions are not unique, it follows that a single development $\phi(x)$ may be derived from an infinite number of different functions $f(x)$. As an example let us take the function

$$f(x)=\sum_0^\infty{}^\nu \frac{c_\nu}{1+a^\nu x},$$

where a is a fixed real quantity > 1. Putting all the A's equal to unity our system (1) becomes

$$\sum c_\nu - 1 = 0,\ \sum c_\nu a^\nu - 1^\nu = 0,\ \ldots,\ \sum c_\nu a^{n\nu} - 1^{n\nu} = 0.$$

---

[1] The correspondence of this latter kind has been pointed out by Borel in the above-mentioned thesis.

[2] Bulletin de la Societé Mathématique de France, Tome VIII.

In order to solve this system we proceed by Poincaré's method; we form a function whose zeros are the points -1, $+1$, a, a^2, ..., *viz.*:

$$F(x) = (1-x)(1+x)\prod_{\nu=1}^{\nu=\infty}\left(1-\frac{x}{a^\nu}\right),$$

a function of genus zero, since $\sum \frac{1}{a^\nu}$ is convergent. We now draw an infinite system of concentric circles C, C_ν, C_1, C_2, ..., C_μ, ..., having the origin as center and such that no circle passes through a zero-point of the function $F(x)$. If now the integral

$$I_{\nu n} = \int_{C_\nu} \frac{x^n dx}{F(x)}$$

taken around a circle C_ν constantly approaches zero, whatever be the value of n, as the circle of integration becomes indefinitely large, it is evident that we must have

$$(2)\quad \sum_0^\infty {}_\nu B_\nu a^{n\nu} - B(1)^n = 0,\ n = 0,\ 1,\ \ldots,\ \infty,$$

where the B's are the residues of the function $\frac{1}{F(x)}$ with respect to the poles -1, 1, a, a^2, ...

These residues are

$$B = \frac{1}{F'(-1)},\ B_0 = \frac{1}{F'(1)},\ B_1 = \frac{1}{F'(a)},\ \ldots,\ B_\nu = \frac{1}{F'(a\ )},$$

and we readily find the following values for the B's,

$$B = \frac{1}{\prod_{\nu=0}^{\nu=\infty}\left(1+\frac{1}{a^\nu}\right)},\quad B_0 = \frac{-1}{2\prod_{\nu=1}^{\nu=\infty}\left(1-\frac{1}{a^\nu}\right)},\ \ldots,$$

$$B_r = \frac{-a^r}{(1-a_r)\prod_{\nu=f}^{\nu=\infty}(r)\left(1-\frac{a^r}{a^\nu}\right)}$$

where the product sign $\Pi_{(r)}$ means that the factor in which $\nu = r$ is absent. These residues give at once the required solutions of our system. In fact, it is only necessary to put

$$C_0 = \frac{B_0}{B}, \; C_1 = \frac{B_1}{B}, \; \ldots, \; C_\nu = \frac{B_\nu}{B}, \; \ldots$$

in order to satisfy the system (1). It remains now to prove that the integral

$$I_{\nu n} = \int_{C_\nu} \frac{x^n dx}{F(x)}$$

approaches zero for $\nu = \infty$. To do this we put

$$F(x) = (1-x)(1+x)\prod_{\nu=1}^{\nu=\infty}\left(1 - \frac{x}{a^\nu}\right) = (1-x)(1+x)\Phi(x).$$

We have

$$|I_{\nu n}| = \left|\int \frac{x^n dx}{(1-x)(1+x)\Phi(x)}\right| < \int \frac{|x^n|\,|dx|}{|(1-x)\;(1+x|\,|\Phi(x)|}$$

$$< \frac{1}{k}\int \frac{|x^n|\;|dx|}{|\Phi(x)|},$$

where k is the smallest value that $|1-x|\,|1+x|$ can have on C_ν. We have thus

$$|I_{\nu n}| < \frac{1}{|1-\rho_\nu^2|}\int \frac{|x^n|\,|dx|}{M_\nu},$$

where M_ν is the smallest value of $|\Phi(x)|$ on C_ν. But putting

$$J_{\nu n} = \int_{C_\nu} \frac{|x^n|\,|dx|}{M_\nu} = \frac{2\pi\rho_\nu^{n+1}}{M_\nu},$$

where ρ_ν is the radius of C_ν, we have

$$|I_{\nu n}| = \frac{1}{|1-\rho_\nu^2|} J_{\nu n}.$$

Let us now suppose that the circles C_ν are drawn in such a way as to satisfy the equation

$$\rho_{\nu+1} = a\rho_\nu,$$

which evidently can be done without ever having any circle pass through a zero of $F(x)$. We have also

$$\Phi(ax) = (1 - x)\Phi(x),$$

so that we get

$$J_{\nu+1,n} = \int_{C_{\nu+1}} \frac{|x^n||dx|}{M_{\nu+1}} = \frac{1}{|1-\rho_\nu|} \frac{2\pi\rho_\nu^{n+1}a^{n+1}}{M_\nu} = \frac{a^{n+1}J_{\nu n}}{|1-\rho_\nu|},$$

which gives at once

$$\frac{J_{\nu+1,n}}{J_{\nu,n}} = \frac{a^{n+1}}{|1-\rho_\nu|} = 0 \text{ for } \nu = \infty .$$

This equation shows that $J_{\nu,n}$ must approach zero, when ν becomes infinite and therefore also $I_{\nu,n}$,—q. e. d.

We have thus arrived at the following simple, BUT FALSE, expansion of $f(x)$,

$$\phi(x) = 1 - x + x^2 - \ldots,$$

which is convergent within the unit circle. The solutions of the system (1) are not unique, in fact, the series

$$c_0 a^0,\ c_1 a,\ c_2 a^2,\ \ldots,\ c_\nu a^\nu,\ -1$$

are equally well solutions of (1), and more generally, $c_0 a^{0P}$, $c_1 a^P$, $\ldots$, $c_\nu a^{\nu P}$, $\ldots$, $(-1)^P$, are solutions of the system; but to all the different functions $f(x)$ that may thus be constructed there corresponds only one expansion $\phi(x)$.

II.

We shall now take up the study of a more general type of functions analogous to the one discussed above. Let

$$(1) \qquad f(x) = \sum_{0}^{\infty}{}_{\nu} \frac{c_\nu}{a_\nu - x}$$

in which $\Sigma|c_\nu|$ is an absolutely convergent series, and we shall suppose $a_0, a_1, \ldots, a_\nu, \ldots$ to be some enumerable ensemble having at least one point-limit not belonging to the ensemble itself, or, to use Cantor's definition, our ensemble is not perfect. We shall further suppose the points a_ν to be condensed at most on lines and not in an area. Let a_0 be some point of the ensemble and let a circle with radius R and center x_0 be drawn through this point and such that no other point of the ensemble lies within the circle.[1] We shall prove the following theorem due to Poincaré:[2]

The series $f(x)$ represents inside of the circle of radius R, center $x_0 = 0$, a monogenic analytic function, and therefore developable in a powerseries convergent for $|x| < R$ and divergent for $|x| \overline{\gtrless} R$. We shall follow Goursat's exposition of the proof.[3]

We have by hypothesis

$$|a_0| = \mathrm{R}, \ \cdots, \ |a_\nu| > \mathrm{R}, \ \nu > 0.$$

If we give to x an absolute value $\nu < \mathrm{R}$, we may develop each term of $f(x)$ in a convergent powerseries

$$f_\nu(x) = \frac{c_\nu}{a_\nu - x} = \frac{c_\nu}{a_\nu} + \frac{c_\nu}{a^{\nu 2}} x + \frac{c_\nu}{a_\nu{}^3} x^2 + \cdots + \frac{c_\nu}{a_\nu{}^{n+1}} x^n + \cdots,$$

$\kappa = 0, 1, 2, \ldots.$ But $f(x)$ considered as the sum of all these series is finite and less than $\frac{1}{r - \mathrm{R}} \sum |c_\nu|$, and therefore the order in which we sum the series is indifferent. Adding then by vertical columns we get

$$(2) \qquad f(x) = \mathrm{A}_0 + \mathrm{A}_1 x + \cdots + \mathrm{A}_n x^n + \cdots,$$

where

$$\mathrm{A}_0 = \sum_0^\infty{}_\nu \frac{c_\nu}{a\nu}, \quad \mathrm{A}_1 = \sum_0^\infty{}_\nu \frac{c_\nu}{a_\nu{}^2}, \quad \ldots, \quad \mathrm{A}_\nu = \sum_0^\infty{}_\nu \frac{c_\nu}{a_\nu{}^{n+1}}, \quad \cdots$$

$f(x)$ is therefore monogenic and analytic within the circle C of

---

[1] An ensemble is said to be condensed in an interval $\alpha, \cdots, \delta$, if an interval $\beta, \cdots, \gamma$, as small as we please, contained in $\alpha, \cdots, \delta$, contains points of the ensemble. See article by G. Cantor, Acta Math., Vol. 2, p. 351.

[2] Acta Soc. Fenn., Tome XII (1883), pp. 341–350.

[3] Sur les Fonctions à Espaces Lacunaires, Bulletin des Sciences Math., 1887, 2ieme Serie, Tome XI, p. 109.

radius R, having its center at the point $x_0 = 0$. Further, the series (2) is divergent if $|x| \overline{\overline{>}} R$. In fact, since $\sum |c_\nu|$ is convergent it is possible to find a number n such that we shall have

$$\sum_{n+1}^{\infty} |c_\nu| < \theta |c_0|, \quad 0 < \theta < 1.$$

We may now decompose $f(x)$ into two parts

$$f(x) = f_1(x) + f_2(x),$$

where

$$f_1(x) = \sum_{1}^{n} \frac{c_\nu}{a_\nu - x}, \quad f_2(x) = \frac{c_\nu}{a_\nu - x} + \sum_{n+1}^{\infty} \frac{c_\nu}{a_\nu - x}.$$

The rational function $f_1(x)$ has poles of mod. $>$ R and may therefore be developed in a series going according to ascending powers of x, convergent for $|x| = R' > R$. As to $f_2(x)$ we have

(3) $$f_2(x) = B_0 + B_1 x + \ldots + B_n x^n + \ldots,$$

where

$$B_n = \frac{c_\nu}{a_0^{n+1}} + \frac{c_{n+1}}{a_{\nu+1}^{n+1}} + \frac{c_{n+2}}{a_{\nu+2}^{n+1}} + \ldots,$$

that is

$$B_n = \frac{1}{a_0^n} \left[\frac{c_0}{a_0} + \frac{1}{a_0} \sum_{n+1}^{\infty} \nu\, c_\nu \left(\frac{a_0}{a_\nu} \right)^{n+1} \right];$$

but we have by hypothesis

$$\frac{a_0}{a_\nu} < 1,$$

and therefore $\left| \sum c_\nu \left(\frac{a_0}{a_\nu} \right)^{n+1} \right|$ will by condition be made less than $\theta |c_0|$, so that we have

$$\theta \left| \frac{c_0}{a_0} \right| > B_n a_0^n < (\theta + 1) \left| \frac{c_0}{a_0} \right|.$$

We may now put the general term of $f_2(x)$ into the form

$$B^n a_0^n \left(\frac{x}{a_0} \right)^n.$$

which shows that the series (3) is divergent if $|x| \overline{\gtrless} a_0 \overline{\gtrless} R$. The two series $f_1(x)$ and $f_2(x)$ have therefore the circle of convergence C, — q. e. d. Suppose now that the points a_ν are condensed along certain lines or arcs of curves. We have then a line of *essential singularities*. Suppose also that our curve L admits of a normal being drawn to it at a point a_0; by the theorem just proved the function $f(x)$ is uniform and analytic within the circle C″ having its center x_0'' on the normal drawn to the point a_0, (Fig. 1), and containing no isolated poles, if

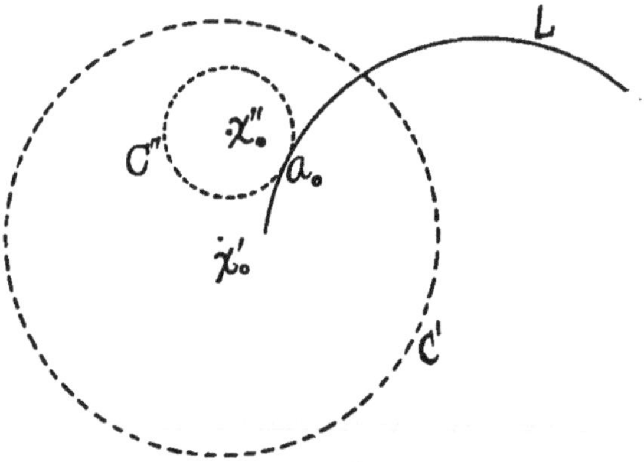

Fig. 1.

such exist. It is moreover evident that any circle whatever, cutting off a part of L cannot be a region of convergence for $f(x)$; for, if so, let x_0' be the center of such a circle C′. Draw now a circle C″ with center x_0'' on the normal to some point a_0 on L and entirely within the larger circle. Since $f(x)$ is an analytic function inside of C′ by supposition, we may develop it at x_0'' in a powerseries $P(x - x_0'')$ having a radius of convergence greaterthan $|x - x_0''|$ which is contrary to the theorem just proved.

The question now naturally presents itself, what happens if the point a_0 is a point-limit not belonging to the ensemble? That such a point differs somewhat from a point of the ensemble itself we have already had occasion to observe in the case of functions of the type $f(x) = \sum \frac{c_\nu}{1 - a_\nu x}$ discussed in the first part of this paper. That $f(x)$, as far as the analytical character of the function is concerned, behaves at a_0 exactly as if this point were an essentially singular point, is evident from the fact that at such a point the function is not expansible in a convergent powerseries, there being an infinite number of poles in the neighborhood of a_0. It may however very well happen, and we shall show this later, that the function, together with all its derivatives, remain finite at such points. As to the corresponding powerseries two things may happen, either is it divergent, in which case it is seen that the point does not differ from an essentially singular point, or it is convergent, having a certain finite or infinite radius of convergence. In this case, however, the powerseries no longer represents the function $f(x)$, which now has ceased to be monogenic and analytic, although it is still finite at the point. We proved the existence of such functions in the first part of this paper.

After these preliminary remarks we shall study a somewhat more special form of functions of the type

$$f(x)=\sum_{0}^{\infty}{}_{\nu} \frac{c_\nu}{x-a_\nu},$$

which we considered above in general. We shall first take the case in which the a's, as well as their point-limits, are condensed on lines or curves. Such lines we shall call *essentially singular lines of the first category*. If point-limits of the a's are condensed on a line in such a way that no point of the ensemble a_ν is situated on it, we shall call such a line *an essentially singular line of the second category*. Functions having such singularities present somewhat different features, as has already been indicated and will be discussed in a subsequent part.

Consider the ensemble

$$a_\nu = e^{2\pi i\nu s}, \quad \nu = 1, 2, \ldots, \infty,$$

where s is an incommensurable fraction. This ensemble is condensed on the unit circle.[1] Let us now put

$$f(x)=\sum_{0}^{\infty}{}_{\nu} \frac{1}{\nu!} \frac{1}{x-e^{\nu i}}$$

[1] I have not been able to find a short and rigorous proof of this. The proof I have found is too lengthy to be given here.

which is derived from (1) by putting $c_\nu = \frac{1}{\nu!}$ and $s = \frac{1}{2\pi}$. This function is uniform and analytic within the unit circle and therefore developable in a powerseries P(x), but does not admit of being continued beyond the circumference of the circle. Now since $f(x)$ is also finite and continuous outside of the unit circle, we may develop it in a powerseries $P\left(\frac{1}{x}\right)$. Further, since an analytic function is defined only in that part of the plane which can be reached by analytic continuation, $f(x)$ represents different analytic functions within and outside of the unit circle, or, as we say, one function is lacunary for the part of the plane in which the other exists.

The respective developments are

$$f_1(x) = -\sum_{n=0}^{n=\infty} x^n . e e^{-(n+1)i}, \quad |x| < 1,$$

$$f_2(x) = \sum_{n=0}^{n=\infty} \frac{e e^{ni}}{x^{n+1}}, \quad |x| > 1.$$

As a more general case consider the function

$$F(x) = \sum_0^\infty {}_\nu \frac{1}{\nu!} \left[\frac{1}{x - r_1 e^{\nu i}} + \frac{1}{x - r_2 e^{\nu i}}\right],$$

admitting the two concentric circles C_{r_1} and C_{r_2} of radii r_1 and r_2 respectively as essentially singular lines. F(x) may now be represented by the series

$$f_1(x) = \quad -\sum_0^\infty {}_n \left(\frac{1}{r_1^{n+1}} + \frac{1}{r_2^{n+1}}\right) e e^{-(n+1)i} . x^n, \quad |x| < r_1,$$

$$f_2(x) = \quad \sum_0^\infty {}_n (r_1^n + r_2^n) e e^{ni} \left(\frac{1}{x}\right)^{n+1}, \quad |x| > r_2.$$

$$f_3(x) = \sum_0^\infty {}_n \frac{1}{r_1} \left(\frac{r_1}{x}\right)^{n+1} e e^{ni} + \sum_0^\infty \left(\frac{x}{r_2}\right)^n e e^{-(n+1)i}, \quad r_1 < |x| > r_2.$$

The first and second developments hold in the region inside C_{r_1} and outside C_{r_2} respectively, while the third holds in the circle-ring formed by C_{r_1} and C_{r_2}; this last development is nothing but Laurent's series.

A still more general function of this kind is

$$F(x)=\sum_{\mu=1}^{\mu=\infty}\sum_{\nu=1}^{\nu=\infty}\frac{c_{\mu\nu}}{r_\mu e^{\nu i}-x},$$

where $\sum\sum c_{\mu\nu}$ is an absolute convergent series. This function has an infinite number of concentric circles C_{r_μ} with radii r_1, r_2, ..., r_μ as lines of singularities. Consider now the circle-ring formed by two concentric circles C_{r_p} and $C_{r_{p+1}}$. We may represent $F(x)$ in this ring by the sum of two powerseries $P(x)$ and $P\left(\frac{1}{x}\right)$ in the same way as before, and we find

$$F(x)=\sum A_\lambda x^\lambda+\sum B_\lambda\left(\frac{1}{x}\right)^\lambda,$$

where

$$B_\lambda=-\sum_{p+1}^{\infty}{}_\mu\sum_{0}^{\infty}{}_\nu\, c_{\mu\nu}(r_\mu e^{\nu i})^{\lambda-1},\quad A_\lambda=\sum_{0}^{p}{}_\mu\sum_{0}^{\infty}{}_\nu\frac{c_{\nu\mu}}{\{r_\mu e^{\nu i}\}^{\lambda+1}}$$

Let us now make the transformation

$$z=\frac{1}{2}\left(x+\frac{1}{x}\right),$$

which transforms the x-plane into the z-plane in such a way that the circles having their origin as center are transformed into confocal ellipses having the points $+1$ and -1 as foci; in particular the unit-circle is transformed into the line $\overline{+1-1}$. Since to a given value of z there corresponds two values of x, we must have two sheets in the z-plane, in order to represent all the values of x in it. Further, the space outside of the unit-circle is transformed into the upper sheet, while the space inside is transformed into the lower sheet, the two sheets having connection all along the line $\overline{+1-1}$.

Suppose now we transform the function

$$f(x)=\sum_{0}^{\infty}{}_{\nu}\frac{c}{e^{\nu i}-x}$$

by the above transformation, we get

$$\Phi(z)=\sum\frac{c_{\nu}}{e^{\nu i}-(z+\sqrt{z^{2}-1})},$$

in which the radical has the sign + or —. Now since the unit-circle is an essentially singular line for $f(x)$, the cut $\overline{+1-1}$, which is evidently nothing but a flattened ellipse, is an essentially singular line for $\Phi(z)$, hence, to $f(x)$ defined as a uniform analytic function within the unit-circle we have a corresponding function $\phi(x)$ in which the sign of the radical must be chosen so as to make $|z+\sqrt{z^{2}-1}|<1$, that is, all the values of z in the lower plane; and to $f(x)$ defined in the space outside of the unit-circle there corresponds a function $\phi_{2}(z)$ in which the radical sign must be chosen so as to make $|z+\sqrt{z^{2}-1}|>1$, that is, $\phi_{2}(z)$ exists in the upper sheet. The development $\sum A_{n}x^{n}$ which defines $f(x)$ within the unit-circle is now transformed into a series $\sum A_{n}(z+\sqrt{z^{2}-1})^{n}$, which is convergent in the lower sheet for all values of z with the exception of the values of z along the cut $\overline{1-1}$.

In the same way, to the development $\sum B_{n}\left(\frac{1}{x}\right)^{n}$, which defines $f(x)$ outside of the unit-circle corresponds a development $\sum B_{n}(z+\sqrt{z^{2}-1})^{-n}$, which is a convergent in the upper sheet outside of the cut. We are thus able to develop the function in the neighborhood of a rectilinear essentially singular line.

Consider now the function

$$f(x)=\sum_{1}^{\infty}{}_{\kappa}\sum_{1}^{\infty}{}_{\nu}\frac{c_{\kappa\nu}\left(k-\frac{1}{k}\right)e^{\nu i}}{(x-ke^{\nu i})\left(x-\frac{e^{\nu i}}{k+1}\right)}$$

which has all the concentric circles with radii 1, 2, ..., ∞, $\frac{1}{2}, \frac{1}{3}, \frac{1}{4}, \ldots, \frac{1}{\infty}$, as essentially singular lines. In the z-plane we get confocal ellipses, situated in such a way that an ellipse in one sheet coincides with an ellipse in the other. Suppose now we develop $f(x)$ in the ring formed by two circles, say the unit-circle and the next one interior to it; we find for the corresponding development in the upper sheet

$$\Phi(z) = \sum_1^\infty A_n(z + \sqrt{z^2 - 1})^n + \sum_1^\infty B_n(z + \sqrt{z^2 - 1})^{-n} + A_0,$$

which series converges for all points included by the cut and the nearest ellipse, provided the sign of the radical be so chosen as to make $|z + \sqrt{z^2 - 1}| > 1$. A similar development holds for the lower sheet in the region coincident with the region in the upper sheet. Let us now put $\Phi(z)$ in the form

$$\Phi(z) = \sum_1^\infty {}^n A_n\left[(z + \sqrt{z^2 - 1})^n - (z - \sqrt{z^2 - 1})^n\right]$$
$$+ \sum_1^\infty {}^n \frac{C_n}{(z + \sqrt{z^2 - 1})^n} + A_0,$$

where $C_n = B_n + A_n$. But $\sum \frac{C_n}{(z + \sqrt{z^2 - 1})^n}$ is convergent outside of cut and the part $\sum A_n\,[(z + \sqrt{z^2 - 1})^n - (z - \sqrt{z^2 - 1})^n]$ has no cut at all, since it converges for all values of z on it and vanishes for $z = \pm 1$. We may therefore put

$$\Phi(z) - \phi(z) = \psi_1(z),$$

where $\psi_1(z)$ is the part having no cut and where $\phi(z)$ is put for the series $\sum \frac{C_n}{(z - \sqrt{z^2 - 1})^n} + A_0$. The analytical continuation of $\psi_1(z)$ from one sheet into another is thus made possible; in fact $\psi_1(z)$ is convergent in the part of the z-plane included by the two coincident ellipses in the upper and lower sheet. Operating on $\psi_1(z)$ exactly as we did with $\phi(z)$ we get

$$\Phi(z) - \phi(z) - \phi_1(z) = \psi_2(z),$$

where $\psi_2(z)$ has no cut and none of the two ellipses nearest to it as essentially singular lines. Continuing in this way we finally arrive at the identity

$$\Phi(z) - \phi(z) - \phi_1(z) - \phi_2(z) - \cdots - \phi_k(z) = \psi_{k+1}(z),$$

where ψ_{k+1} is a function which is holomorphic in the whole region included between $k+1$ ellipses in the upper sheet and the corresponding ellipses exactly below it. The form of the function $\psi_{k+1}(z)$ is easily obtained. We only need to develop $f(x)$ by Laurent's theorem in the region between the kth and $k+1$st ellipse. We obtain

$$f(x) = \sum_{1}^{\infty} A_n' x^n + \sum_{1}^{\infty} B^{n'} \left(\frac{1}{x}\right)^n + A_0;$$

the first series on the right-hand side is convergent for all values of x, satisfying the condition $|x| < k+1$. We now write as before

$$f(x) = \sum A^{n'}\left(x_n - \frac{1}{x^n}\right) + \sum \frac{B_n' + A_n'}{x^n} + A_0',$$

where the first series is convergent, if $k+1 > |x| > \frac{1}{k+1}$, and the second series is convergent outside of the circle C_k. Hence we may put

$$f(x) - \phi_1(x) = \psi_{\substack{1 \\ k+1}}(x) = \sum A_n'\left(x^n - \frac{1}{x^n}\right),$$

or, transforming into the z-plane,

$$\Phi(z) - \phi(z) = \psi_{k+1}(z) = \sum A_n'[(z+\sqrt{z^2+1})^n - (z-\sqrt{z^2-1})^n],$$

where ψ_{k+1} is a holomorphic function in the region between the ellipses specified above and vanishes at the two branch-points.

We have thus seen that the function $f(x)$, or its transform $\phi(z)$, belongs to that class of functions which can be put in the form

$$f(x) = \phi(x) + \psi(z),$$

where $\phi(x)$ has one essentially singular line C and $\psi(z)$ a uniform function which may or may not have other singular lines, but not the singular line C or any part of it. The difference $f(x) - \phi(x)$ has therefore no essentially singular line C. That this is not always possible is easily seen from an example given by Borel,

$$F(z) = \phi(z) + \phi_1(z) \log z,$$

where $\phi(z)$ has the negative part of the real axis as essentially singular line and $\phi_1(z)$ is a polynomial. Although $F(z)$ is an analytic and *one valued* function outside the cut, it is not possible to continue it across this line by subtracting any term of the right side from the left, in fact, the difference

$$F(z) - \phi(z) = \phi_1(z) \log(z)$$

is no longer uniform; this is due to the fact that $F(z)$ is made uniform in an artificial way by providing $\phi(z)$ with a singular line, which prevents the variable z to describe a complete path around the origin.

Another interesting question here suggests itself: Is it possible to pass continuously with a function $f(x)$ across any of the ellipses, say the cut $1 - 1$, that is, from one sheet to another? It is evident that as long as we define $f(z)$ as a uniform analytic function in Weierstrass' sense no such passage is possible, since the path of the variable is not restricted at all, provided it remains within the circle of convergence. But suppose we restrict the variable to move along certain paths; in that case we may formulate the question thus: Is it possible to pass across an essentially singular line without the function suffering a break of continuity? That this is possible has been proved in a theorem by Borel, which we shall state here without proof.[1]

Consider the series

[1] See Borel's thesis, "Sur Quelques Points, etc." Paris, 1894.

$$\sum \frac{A_n}{(z - a_n)^{m_n}},$$

in which A_n is an ensemble of points that may be condensed along lines, or even in areas, and m_n has an upper finite limit n. Suppose now that a convergent series of positive terms Σu_n exists such that the series

$$\sum \frac{|A_n|}{u_n^{m_n}}$$

is convergent. In order to effect this it is only necessary and sufficient that the series of positive terms

$$\sum \sqrt[m_n + 1]{|A_n|}$$

shall be convergent. Having now given two points, P and Q, which do not coincide with a point A_n, nor with any point-limit of the ensemble, Borel proves that *it is always possible to join these points by a continuous line* γ, *such that the series considered shall be absolutely and uniformly convergent*, *as well as all its derivatives*, *on this line.*

The determination of the line γ must, in general, be effected for each individual function and constitutes the most difficult problem in the application of this important theorem.

It is, however, not difficult to prove that in the case of the function $f(x)$, that we have been dealing with, certain lines through the origin may be chosen as the line γ, and that, moreover, an infinite number of such lines exist. In fact, the possibility of passing through the unit circle without break of continuity will only depend on the possibility of choosing coefficients c_ν in such a way that

$$f(e^{\theta i}) = \sum_{1}^{\infty}{}_{\nu} \frac{c_\nu}{e^{\theta i} - e^{\nu i}}$$

shall be absolutely convergent for one or more values of θ not coinciding with any one of the values $1, 2, 3, \ldots, \infty$. We have

$$\sum_{1}^{\infty} \frac{c_\nu}{e^{\theta i} - e^{\nu i}} = \sum \frac{c_\nu}{\cos\theta - \cos\nu + i(\sin\theta - \sin\nu)}$$

$$= \tfrac{1}{2} \sum \frac{c_\nu}{-\sin\frac{\theta+\nu}{2}\sin\frac{\theta-\nu}{2} + i\left(\cos\frac{\theta+\nu}{2}\sin\frac{\theta-\nu}{2}\right)}$$

$$= -\frac{i}{2} \sum \frac{c_\nu}{\sin\frac{\theta-\nu}{2}\, e^{\frac{\theta+\nu}{2}i}};$$

that is, we must have the series

$$(1) \qquad \sum \frac{c_\nu}{\sin\frac{\theta+\nu}{2}}$$

convergent for one or more values of θ. In the first place, suppose $\theta = 0$ or π, in order that (1) may be convergent it is only necessary to put

$$c_\nu = c_\nu' \sin\nu$$

where c_ν' is a series of quantities satisfying the conditions

$$\lim_{\nu=\infty} \frac{c_\nu'}{c'_{\nu-1}} = 1, \quad \lim_{\nu=\infty} c_\nu = 0.$$

These conditions being satisfied, the series

$$\sum c_\nu' \sin\frac{\nu}{2}, \quad \sum c_\nu' \cos\frac{\nu}{2}$$

are absolutely convergent (See Picard Traité d'An., T. I., p. 231) and our theorem is proved for two values of θ. If then we follow the path along the real axis, we are able to pass continuously through the unit circle at the points ± 1. In general, suppose we choose our coefficients

$$c_\nu = c_\nu' \sin\nu \prod_{\theta=\frac{m}{n}2\pi} a_{mn} \sin\frac{\theta-\nu}{2} \qquad \begin{matrix} n = 3, 4, 5, \dots, \infty \\ m = 1, 2, 3, \dots, n-1 \end{matrix}$$

where $\Sigma c_\nu'$ is an absolutely convergent series. The product

$$\prod_{\theta=\frac{m}{n}2\pi} a_{m,n} \sin\frac{\theta-\nu}{2}$$

will be absolutely convergent for any fixed value of ν if the series $\sum \left(a_{m,n} \sin \frac{\theta - \nu}{2} - 1\right)$ enjoys the same property. Having chosen the coefficients $a_{m,n}$ so as to satisfy this condition, we form the function

$$f(x) = \sum \frac{c_\nu' \sin \nu \prod a_{m,n} \sin \frac{\theta - \nu}{2}}{x - e^{\nu i}},$$

which is absolutely convergent at the points $\theta = \frac{m}{n} 2\pi$. We may therefore pass continuously with $f(x)$ through the circle along lines whose directions at the points of exit pass through the origin and form angles with the real axis equal to any commensurable part of 2π. These lines will in the z-plane be represented by homofocal hyperbolæ leading from one sheet into another.

III.

We shall now consider a class of functions having essentially singular lines of the second category. We defined these as lines on which are condensed point-limits of the ensemble not belonging to the ensemble itself. We said above that on such lines the function may be finite and continuous as well as all its derivatives. Prof. Mittag-Leffler in a note in *Acta Math.*, Vol. 15, called the attention of mathematicians to a function of this nature admitting of no analytic continuation across the unit circle, but possessing the property of remaining finite and continuous on this line as well as all its derivatives. Pringsheim in a paper already mentioned discusses a class of functions which seem to posses a similar property. He starts with the expression

$$f(z) = \sum_0^\infty \frac{c_\nu}{z - a_\nu},$$

where the ensemble a_ν has its point-limits condensed on a line in such a way that all the poles are on one side of it; thus, for in-

stance, we may put $a_\nu = \left(1 + \frac{1}{\nu}\right)e^{\nu i}$, the poles approaching the unit circle in a narrowing spiral. Pringsheim shows that it is always possible to choose coefficients c_ν in such a way as to make $f(z)$ and all its derivitives finite and continuous on the circle. In fact, we have

$$|f(e^{i\theta})| < \sum \frac{c_\nu}{|z| - |a_\nu|} < \sum \frac{|c_\nu|}{|a_\nu| - 1}.$$

In order that this last series may be convergent we must put

$$(1) \qquad |c_\nu| = |c_\nu'| \, (|a_\nu| - 1),$$

where c_ν' are the terms of the convergent series $\Sigma c_\nu'$, and in general, since

$$f^{(n)}(z) = (-1)^n n! \sum \frac{c_\nu}{(z - a_\nu)^{n+1}},$$

we have

$$|f^{(n)}(e^{i\theta})| < n! \sum \frac{|c_\nu|}{|e^{i\theta} - a_\nu|^{n+1}} < n! \sum \frac{|c_\nu|}{|a_\nu - 1|^{n+1}};$$

that is, we must put

$$(2) \qquad |c_\nu| = |c_\nu'| \, (a_\nu - 1)^{n+1},$$

where $\Sigma c_\nu'$ is a convergent series. This condition being satisfied the condition (1) is also satisfied. In order to accomplish our purpose we only need to put $n = \nu$ in (2) so as to insure the convergence of $\sum \frac{c_\nu}{|a_\nu - 1|^{n+1}}$ for *all* values of n; we therefore put

$$(3) \qquad |c_\nu| = |c_\nu'| \, (a_\nu - 1)^{\nu+1},$$

so that the function $f(z)$ and all its derivatives up to any order as high as we please will be absolutely convergent on the unit circle, if the c_ν's satisfy the condition (3). Thus, suppose we put $a_\nu = \left(1 + \frac{1}{\nu}\right)e^{\nu i}$; we also make $|c_\nu| = \frac{1}{\nu^2}\left(1 - 1 - \frac{1}{\nu}\right)^{\nu+1}$ $= \frac{1}{\nu^2} \frac{1}{\nu^{\nu+1}}$ and we have

$$|f^{(n)}(e^{i\theta})| < n!\sum \frac{1}{\nu^2}\,\frac{\dfrac{1}{\nu^{\nu+1}}}{\dfrac{1}{\nu^{n+1}}} = n!\sum \frac{1}{\nu^2}\,\frac{1}{\nu^{\nu-n}},$$

a convergent series for all values of n as large as we please.

An objection may be raised against some of Pringsheim's conclusions, and it is indeed instructive to notice how very careful we must be in drawing conclusions from an apparent property of a function represented in the form of an arithmetical expression. Pringsheim reasons thus: a function

$$f(z) = \sum \frac{c_\nu}{z - a_\nu}$$

with the unit circle as singular line admits of no analytical continuation across the unit circle, since in the neighborhood of any

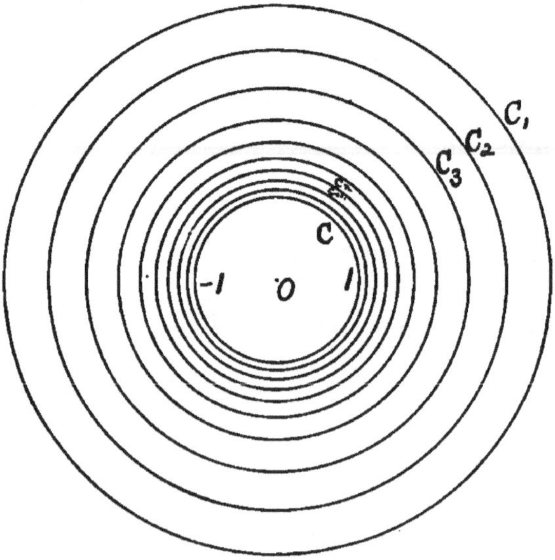

point on the circle there is an infinite number of poles of $f(z)$. The function $f(z)$ is therefore two distinct functions within and outside of the unit circle. *But this is not necessarily true.* In fact, we may imagine the existence of a function of the same form and property as $f(z)$ for which an analytical continuation is possible. That such a function can be formed has been shown by Prof. Phragmén, who proceeds thus:[1]

Let $f(u)$ be a given analytic function, holomorphic within a region A. Choose within A an infinite number of circles C_n, such that C_{n+1} lies entirely within C_n and let C_n converge toward a given circle C, which may be represented by the unit circle, as n increases indefinitely. Choose positive quantities ϵ_1, $\epsilon_2, \ldots, \epsilon_\nu$ such that $\lim_{\nu=\infty} \epsilon_\nu = 0$. Since $f(u)$ is holomorphic within A, we may represent this function in the region within C_1 by Cauchy's integral

$$f(z) = \int_{C_1} \frac{f(u)}{u-z}\, du$$

taken along C_1. We may therefore choose a number of points u_ν' on C_1 and corresponding values $A_\nu' = f(u_\nu')(u'_{\nu+1} - u_\nu')$ such that

$$f_1(z) = \sum \frac{a_\nu'}{u_\nu' - z}$$

everywhere within C_1 differs from $f(z)$ by less than ϵ_1, that is

$$| f(z) - f_1(z) | < \varepsilon.$$

In the same way, since $f(z) - f_1(z)$ may likewise be represented by a Cauchy's integral taken around C_2, we may form the expression

$$f_2(z) = \sum \frac{A_\nu^{(2)}}{u_\nu^{(2)} - z}$$

whose poles lie on C_2 and satisfy the inequality

$$|f(z) - f_1(z) - f_2(z)| < \epsilon_2.$$

Continuing in this way we may form new expressions $f_3(z)$, etc., and we have finally

$$(1) \qquad f(z) = \sum_1^\infty {}_\nu f_\nu(z),$$

[1] Phragmén's method of forming this function has been communicated to me in a letter from Prof. Bjerkness of Kristiania University.

an expression which is uniformly convergent within C and represents $f(z)$ in that region.

It would be easy to choose our expressions such that it converges even outside of C, leaving out the poles of course. If σ_1, σ_2, ..., σ_ν, ..., be a series of positive quantities chosen in such a way that $\Sigma\sigma_\nu$ converges, we may, as is seen by comparing with Cauchy's integral, choose $f_1(z), f_2(z)$, ..., in such a way that

$$|f_1(z)| < \sigma_1 \text{ outside of } C_1$$
$$|f(z)| < \sigma_2 \text{ outside of } C_2$$
$$\vdots$$
$$|f_{\nu+1}(z)| < \sigma_\nu \text{ outside of } C_\nu$$

and from these inequalities it follows that the expression (1) will be uniformly convergent outside of C. This function then possesses the same property as Pringsheim's, but we know that, since $f(z)$ within C is identical with the given holomorphic function, it can be continued beyond this line. This seeming paradox may find an explanation in the following way. Suppose we have a function holomorphic in a given region which we may suppose to be the unit circle around the origin, the poles all being outside of it, and we may suppose these to be forming singular lines such that $\lim_{\nu=\infty} a_\nu \geqq 1$. Let this function be represented by the series

$$f(z) = \sum \frac{c_\nu}{a_\nu - z},$$

all the poles being of multiplicity 1. Suppose further that within the region mentioned the following equality holds

$$f(z) = \phi(z),$$

where $\phi(z)$ is a given function holomorphic in a region including at least the unit circle. This function may be put in the form

$$\phi(z) = A_0 + A_1 z + A_2 z^2 + \cdots + A_n z^n + \cdots,$$

convergent for $|z| < R'$, $R' \geqq R$.

Since $f(z)$ is holomorphic within the region $|z| < 1$, it may be developed in a convergent powerseries

$$f(z) = \sum \frac{c_\nu}{a_\nu} + \sum \frac{c_\nu}{a_\nu^2} . z + \dots + \sum \frac{c_\nu}{a_\nu^n} z^n + \dots .$$

Now since $f(z)$ is to be identical with $\phi(z)$ for all values of z, such that $|z| < R'$, we must have

$$(2) \qquad \begin{aligned} \sum \frac{c_\nu}{a_\nu} &= A_0, \\ \sum \frac{c_\nu}{a_\nu^2} &= A_1, \\ &\vdots \\ \sum \frac{c_\nu}{a_\nu^n} &= A_n, \\ &\vdots \end{aligned}$$

The identity $f(z) = \phi(z)$ within the region $|z| < R'$ leads therefore to a system of an infinite number of equations with an infinite number of unknowns which must be satisfied by the c's, and these are thus seen to be functions of the a's. Conversely, if solutions of (2) exist, it will be possible to make $f(z)$ equal to any given holomorphic function. Now we do not know in general how to solve such a system. Poincaré's method fails of course, since here $\lim\limits_{\nu=\infty} \frac{1}{a_\nu}$ equals some finite quantity, while it ought to be equal to ∞. That in certain cases solutions do exist seems evident when we consider Phragmén's example, $f(z) = \sum f_n(z)$, where $f_n(z) = \sum \frac{A_\nu^{(n)}}{u_\nu^{(n)} - z}$, which, when developed, gives rise to a system of equations similar to (2). If then the coefficients are not solutions of (2), we may be sure that no ambiguity can arise, so that, when we speak of developing $f(z)$ across the unit circle, we mean $f(z)$ and no function that can be so continued and which is identical with it for any particular region. *The possibility that $f(z)$ may represent any given* *function in one part of the plane and a transcendental function in another has been overlooked by Pringsheim.* There can be no doubt, however, that the functions which he discusses do not possess this property, as it is evident that a series of quantities c_ν, only subject to the condition $\Sigma c_\nu =$ a convergent series, do not in general satisfy the system (2).

Another method of constructing functions possessing the property of being finite and continuous, as well as all its derivatives, on the unit circle has been given by Borel who starts with the function

$$\phi(z) = \sum b_\nu x^{c_\nu},$$

where the exponents c_ν fulfill the condition

$$\frac{c_\nu - c_{\nu-1}}{c_\nu} > N.$$

N being a fixed quantity. The unit circle being a singular line for $f(z)$ (See Hadamar's paper, Lionville, Fourth series, T. VII, p. 115), we may always choose coefficients b_ν in such a way as to render $\phi(z)$ and all its derivatives finite and continuous on the circle. The function constructed by Fredholm belongs to this type. If we transform such a function by means of the well-known transformation $z = e^{i\theta}$ and take the real part of it, we obtain a new function $f(\theta)$ of a real variable to which Taylor's development cannot be applied in spite of the fact that $f(\theta)$ and all of its derivatives are finite and continuous for all values of the variable.[1] The important connection that thus is seen to exist between certain functions of an imaginary variable and functions of a real variable which are not developable at any point in a Taylor's series may perhaps justify the search for other transcendentals possessing the same property as the type given by Pringsheim and Borel, but of a more general character, and the author has, he believes, succeeded in adding a new and interesting class of transcendentals to those given above.

We shall call a point an *essentially singular point of the*

[1] For proof of this theorem see article by Pringsheim.

second category, if it is a point-limit of poles of a function, and we shall use the term *essentially singular point of the first category* for points a such that no finite power ω exists, which renders $\lim_{z=a} (z-a)^{\omega} f(z)$ uniformly finite whatever be the path along which z tends to a. If points of the second category only are condensed along lines, these will become *singular lines of the second category* by our former definition (p. 17) and if points of the first and second categories are condensed on lines, these will become *lines of the first category*. A point of the second category may be a point-limit of poles *as well as of essentially singular points.* We presuppose, as before, that no point-limit belongs to the ensemble from which it is derived. It is also evident that whether the singular line be of the second or first category, it will affect the function in the same way; *viz.*, prevent any analytic continuation beyond it, provided, of course, the line be closed.

Consider now the expression

$$(1) \qquad F(z) = \prod_{\nu=0}^{\nu=\infty} \frac{z-(1+a_\nu)e^{2\pi i\nu a}}{z-(1+b_\nu)e^{2\pi i\nu a}},$$

where $\lim_{\nu=\infty} a_\nu = 0$, $\lim_{\nu=\infty} b_\nu = 0$, and a an incommensurable number, which without loss of generality may be put equal to $\frac{1}{2\pi}$. The quantities a_ν and b_ν being supposed positive, the ensembles $(1+a_\nu)e^{\nu i}$, $(1+b_\nu)e^{\nu i}$ will represent points outside the unit circle approaching it asymptotically as ν increases indefinitely. The points $e^{\nu i}$ which are condensed on the unit circle will therefore be point-limits of the given ensembles. In order that the product

$$(2) \qquad \prod \frac{z-(1+a_\nu)e^{\nu i}}{z-(1+b_\nu)e^{\nu i}},$$

shall be uniformly convergent it is necessary and sufficient that the series

$$(3) \qquad \sum |a_\nu - b^\nu|$$

be absolutely convergent. The ensembles a_ν and b_ν satisfying this condition, the product (1) will represent a function $F(z)$ which is finite and continuous throughout the whole plane with the exception of the points $(1+b_\nu)e^{\nu i}$ and the points $e^{\nu i}$ on the circle. Within the unit circle $F(z)$ will represent a uniform and analytic function admitting this circle as a singular line. This line plays a double rôle with reference to the function $F(z)$, the points $e^{\nu i}$ being point-limits of poles *as well as of zeros*, and it is evident that, if we draw around any point of it a circle however small, an infinite number of poles and zeros will be situated within the circle. We may call such a line a *doubly singular line of the second category*. As a particular example, suppose we put $a_\nu = \frac{1}{\nu^2}$, $b_\nu = \frac{1}{2\nu^2}$ which make $\sum |a_\nu - b_\nu| = \frac{1}{2}\sum \frac{1}{\nu^2}$ = a convergent series. The corresponding function will be

$$F(z) = \prod_1^\infty \frac{z - \left(1 + \frac{1}{\nu^2}\right)e^{\nu i}}{z - \left(1 + \frac{1}{2\nu^2}\right)e^{\nu i}}$$

and will have the properties described above. An infinite number of ensembles a_ν and b_ν may manifestly be formed satisfying the condition (3). Examples of such ensembles are

$$a_\nu = \frac{1}{e^\nu},\quad b_\nu = \frac{1}{e^\nu - \nu};\quad a_\nu = \frac{1}{2^\nu},\quad b_\nu = \frac{1}{2^{\nu+1}}, \text{ etc.}$$

Remark.—It should be noticed that the form of the ensembles $(1+a_\nu)$, $(1+b_\nu)$ is not as special as might be supposed; in fact, the ensemble e^ν is for our purpose just as appropriate, since it may always be put in the form $1+a_\nu$, a_ν being equal to $\frac{1}{\nu} + \frac{1}{2\nu^2} + \frac{1}{3!\,\nu^3} + \ldots$, and more generally a^{k_ν}, where $|a| > 1$ and $\lim\limits_{\nu=\infty} k_\nu = 0$, may also be put in the same form.

We shall now take up the study of functions like $F(z)$ on the singular line which we shall suppose, as before, to be the unit circle around the origin. Let then

$$F(z) = \prod_{\nu=1}^{\nu=\infty} \frac{z - (1+a_\nu)e^{\nu i}}{z - (1+b_\nu)e^{\nu i}},\quad \lim_{\nu=\infty} a_\nu = 0,\quad \lim_{\nu=\infty} b_\nu = 0,$$

and let also the condition (3) be fulfilled. Putting $F(z)$ into the form

$$F(z) = \prod \left(1 - \frac{(a^\nu - b_\nu)e^{\nu i}}{z - (1 + b_\nu)e^{\nu i}}\right)$$

we see that if $F(z)$ shall be convergent on the unit circle, that is for $|z| = 1$, we must have

$$\sum \frac{(a_\nu - b_\nu)e^{\nu i}}{z - (1 + b_\nu)e^{\nu i}}$$

a convergent series for $|z| = 1$; but we have

$$\sum \frac{|a_\nu - b_\nu|}{|z - (1 + b_\nu)e^{\nu i}|} < \sum \frac{a_\nu - b_\nu}{|b_\nu + 1| - |z|};$$

that is, we must have

$$(4) \qquad \sum \left|\frac{a_\nu - b_\nu}{b_\nu}\right| = \text{an abs. conv. series.}$$

If then this condition is fulfilled, $F(z)$ will be finite and continuous on the circle, the condition (4) including the condition (3). By proper choice of ensembles we may satisfy this condition; we may for example suppose the ensembles a_ν and b_ν written in the form

$$a_\nu = \frac{1}{a_\nu'},\quad b_\nu = \frac{1}{a_\nu' \pm b_\nu'},\quad \lim_{\nu=\infty} a_\nu' = \infty,$$

in which case our condition reduces to the simpler one $\sum \frac{b_\nu'}{a_\nu'}$ = a convergent series.

Examples of such ensembles are

$$a_\nu = \frac{1}{e^\nu},\quad b_\nu = \frac{1}{e^\nu - 1};\quad a_\nu = \frac{1}{2^\nu},\quad b_\nu = \frac{1}{2^\nu - 1},\quad \text{etc.,}$$

or, choosing the form

$$a_\nu = a_\nu' + b_\nu,\quad b_\nu = b_\nu,$$

we have the condition

$$\sum \frac{a_\nu'}{b_\nu} = \text{a convergent series.}$$

We may thus, for instance, put

$$a_\nu = \frac{1}{e^\nu} + \frac{1}{\nu^2},\ b_\nu = \frac{1}{\nu^2};\ a_\nu = \frac{1}{a^\nu} + \frac{1}{\nu^k},\ b_\nu = \frac{1}{\nu^k},\ \text{etc.}$$

$|a|$ being greater than unity and k some position integer > 1.

We have thus proved the theorem.

It is always possible to choose ensembles a_ν and b_ν such that the function

$$F(z) = \prod \frac{z - (1 + a_\nu)e^{\nu i}}{z - (1 + b_\nu)e^{\nu i}}$$

where a_ν and b_ν fulfil the condition $\lim_{\nu=\infty} a_\nu = 0$, $\lim_{\nu=\infty} b_\nu = 0$,

$$\sum \left| \frac{a_\nu - b_\nu}{b_\nu} \right| = \textit{an absolutely convergent series, shall be finite and}$$

continuous within as well as on the unit circle.

A remark concerning the continuity of $F(z)$ on the unit circle is here in place. It is evident that $F(z)$ is not continuous for points on the unit circle in the ordinary sense, since for these points no complete region around any one of them can be formed such that $|F(z + \lambda z) - F(z)| < \epsilon$. This inequality can only be satisfied for a region around the point lying completely within the unit circle having a part of the circle-line as boundary. If the variable moves within this region, or on the boundary of it, the above inequality can be satisfied by taking the region small enough. We may therefore say that $F(z)$ is *inwardly continuous* at all points of the unit circle.

The question now remains, What happens to the derivatives of $F(z)$? Forming the first derivatives we have

$$F'(z) = F(z) \times \sum \left[\frac{1}{z - (1 + a_\nu)e^{\nu i}} - \frac{1}{z - (1 + b_\nu)e^{\nu i}} \right] =$$

$$F(z) \times \sum \frac{(a_\nu - b_\nu)e^{\nu i}}{[z - (1 + a_\nu)e^{\nu i}][z - (1 + b_\nu)e^{\nu i}]},$$

and, since $F(z)$ is already absolutely convergent for $|z| = 1$, in order that $F'(z)$ shall possess the same property, it is necessary and sufficient that

$$\left| \sum \frac{(a_\nu - b_\nu)e^{\nu i}}{[z - (1 + a_\nu)e^{\nu i}][z - (1 + b_\nu)e^{\nu i}]} \right| <$$

$$\sum \frac{|a_\nu - b_\nu|}{[a_\nu + 1 - |z|][b_\nu + 1 - |z|]}$$

shall be absolutely convergent for $|z|=1$; this leads to the condition

$$(5) \qquad \sum \left| \frac{a_\nu - b_\nu}{a_\nu b_\nu} \right| = \text{an abs. conv. series,}$$

which evidently also includes the condition (4). Let us now write the first derivative in the form

$$F'(z) = F(z)\Phi(z);$$

we find for the second derivative

$$F''(z) = F'(z)\Phi(z) + F(z)\Phi'(z).$$

The condition (5) being fulfilled, $F(z)$, $\Phi(z)$ and $F'(z)$ are all finite and continuous on the unit circle, and it is therefore only necessary that $\Phi'(z)$ be absolutely convergent for $|z|=1$ in order to have $F''(z)$ enjoy the same property. This leads again to the following condition

$$(6) \qquad \sum \left[\left(\frac{1}{a_\nu}\right)^2 - \left(\frac{1}{b_\nu}\right)^2 \right] = \text{an abs. conv. series.}$$

If this condition be satisfied, the condition (5) and (4) will manifestly also be satisfied. In general, forming the nth derivative of $F(z)$ we have

$$F^n(z) = F^{n-1}(z)\Phi(z) + (n-1)F^{n-2}(z)\Phi'(z) + \dots + \Phi^{n-1}(z)F(z),$$

and from this formula we see that if the derivatives of order lower than n are absolutely convergent for $|z|=1$, $F^n(z)$ will possess the same property, provided also $\Phi^{n-1}(z)$ be absolutely convergent. But this again leads to the condition

$$(7) \qquad \sum \left[\left(\frac{1}{a_\nu}\right)^n - \left(\frac{1}{b_\nu}\right)^n \right] = \text{an abs. conv. series.}$$

If this condition is satisfied, then will also all the derivatives of order lower than n be finite for $|z|=1$; in fact, if condition (7) is satisfied, then will also the series

$$\sum \left[\left(\frac{1}{a_\nu}\right)^r - \left(\frac{1}{b_\nu}\right)^r \right],$$

where r is any positive integer less than n, be convergent; but this is nothing but the condition that must be fulfilled in order that the rth derivative of F(z) shall be absolutely convergent on the unit circle *q. e. d.* We may therefore say,

The necessary and sufficient condition that $F(z)$ together with all its derivatives up to the nth order, n being a positive integer as high as we please, shall be absolutely convergent on the unit circle is

$$\sum\left[\left(\frac{1}{a_\nu}\right)^n-\left(\frac{1}{b_\nu}\right)^n\right]=\text{an abs. conv. series,}$$

a_ν and b_ν being subject to the conditions $\lim\limits_{\nu=\infty} a_\nu=0$, $\lim\limits_{\nu=\infty} b_\nu=0$.

If then we can find ensembles a_ν and b_ν satisfying the above conditions we shall have solved the problem of constructing a function, admitting the unit circle as doubly singular line and such that the function itself with all its derivatives up to any order as high as we please remain absolutely convergent on this line.

Now I say that

$$a_\nu=\frac{1}{2^\nu},\ b_\nu=\frac{1}{2^\nu-\frac{1}{2^{\nu^2}}}$$

are such ensembles. In fact, we have $\lim\limits_{\nu=\infty} a_\nu=0$, $\lim\limits_{\nu=\infty} b_\nu=0$,

$\sum\left[a_\nu^{-n}-b_\nu^{-n}\right]=\sum\left[2^{n\nu}-\left(2^\nu-\frac{1}{2^{\nu^2}}\right)^n\right]$ and this latter series is easily seen to be absolutely convergent for values of n however large. We have

$$\sum\left[2^{n\nu}-\left(2^\nu-\frac{1}{2^{\nu^2}}\right)^n\right]=\sum\left[\frac{n2^{(\nu-1)n}}{2^{\nu^2}}\cdots-(-1)^n\frac{1}{(2^{\nu^2})^n}\right]$$

$$<n^2\sum\frac{2^{(\nu-1)n}}{2^{\nu^2}}<n^2\sum\frac{2^{\nu n}}{2^{\nu^2+n}}$$

and this last series is convergent for all positive integer values of n however large ;, the function

$$F(z)=\prod_{\nu=1}^{\nu=\infty}\frac{z-\left(1+\frac{1}{2^\nu}\right)e^{\nu i}}{z-\left(1+\frac{1}{2^\nu-\frac{1}{2^{\nu^2}}}\right)e^{\nu i}}$$

will therefore have the required property. The condition (7) reduces to a simpler form if we write the ensembles a_ν and b_ν in the form

$$a_\nu = \frac{1}{a_\nu'},\quad b_\nu = \frac{1}{a_\nu' \pm b_\nu'},$$

and we have

$$\Sigma[a_\nu^{-n} - b_\nu^{-n}] = \Sigma[a_\nu'^n - (a_\nu' \pm b_\nu')^n] < \Sigma[\mp n a_\nu'^{n-1} \pm \cdots]$$
$$< n'\Sigma a_\nu'^{n-1} b_\nu'.$$

This last series must therefore be convergent for values of n however large, and this may obviously be effected by making a_ν' and b_ν' satisfy the condition

(8) $\qquad \Sigma a_\nu'^{\nu-1} b_\nu' =$ an abs. conv. series.

Examples of such ensembles are

$$a_\nu' = \nu^2,\ b_\nu' = \frac{1}{\nu^{2\nu}};\ a_\nu' = 2^\nu,\ b_\nu' = \frac{1}{2^{\nu^2}};\ a_\nu' = a^\nu,\ b_\nu' = \frac{1}{a^{\nu^2}},$$

a being a positive quantity > 1. Again, suppose we put the ensembles in the form

$$a_\nu = \frac{1}{a^{\nu'}},\quad b_\nu = \frac{1}{a^{\nu'}} \pm \frac{1}{b_\nu'}$$

we have

$$\sum\left[a_\nu'^n - \frac{a_\nu'^n}{\left[1 \pm \frac{a_\nu'}{b_\nu'}\right]^n}\right] < \sum\left[a_\nu'^n - a_\nu'^n\left(1 \mp \frac{a_\nu'}{b_\nu'}\right)^n\right] <$$
$$n'\sum a_\nu'^{n+1} b_\nu'^{-1},$$

which last series must be absolutely convergent; that is

(9) $\qquad \Sigma a_\nu'^{\nu+1} b_\nu'^{-1} =$ an abs. conv. series.

Such ensembles are, for instance,

$$a_\nu' = \nu^2,\ b_\nu' = \nu^{2\nu+4};\ a_\nu' = 2^\nu,\ b_\nu' = 2^{\nu^2+2\nu},\ \text{etc.}$$

We have thus proved the following theorem:

It is always possible to choose ensembles a_ν and b_ν in such a way that the function

$$F(z) = \prod \frac{z - (1 + a_\nu)e^{\nu i}}{z - (1 + b_\nu)e^{\nu i}},$$

and all its derivatives up to any order as high as we please shall be finite and continuous within and on the unit circle around the origin, admitting this line as a doubly singular line.

The function $F(z)$ being analytic within the unit circle may be developed in the neighborhood of an ordinary point z' lying in this region in a powerseries $P(z-z')$. If z' be a point on the unit circle, a powerseries, even if convergent, will manifestly cease to represent the function.

The special property of $F(z)$ on the unit circle depends, as we have seen, on the choice of the ensembles a_ν and b_ν. The form of $F(z)$ is, however, rather special, the points of discontinuity of $F(z)$ being simple poles and point-limits of poles. A more general function admitting the unit circle as a doubly singular line, but whose points of discontinuity are essential singularites and point-limits of such, may be constructed as follows:

Consider the expression

$$(10) \qquad \psi(z) = \prod \frac{z-(1+a_\nu)e^{\nu i}}{z-(1+b_\nu)e^{\nu i}} \cdot e^{\Phi_\nu(z)},$$

where

$$\Phi_\nu(z) = \frac{(a_\nu - b_\nu)e^{\nu i}}{z-(1+b_\nu)e^{\nu i}} + \frac{1}{2}\left[\frac{(a_\nu - b_\nu)e^{\nu i}}{z-(1+b_\nu)e^{\nu i}}\right]^2 + \cdots$$
$$+ \frac{1}{\nu-1}\left[\frac{(a_\nu - b_\nu)e^{\nu i}}{z-(1+b_\nu)e^{\nu i}}\right]^{\nu-1}$$

and we shall suppose, as before, lim. $a_\nu = 0$, lim. $b_\nu = 0$. We shall prove that $\psi(z)$ is a holomorphic function everywhere in the plane where $|z-(1+b_\nu)e^{\nu i}| > |a_\nu - b_\nu| > 0$, and in particular within the unit circle. In establishing this proposition we shall adopt, with slight modification, a method of proof applied by Picard to functions of a similar form.[1] Denoting the general factor of $\psi_\nu(z)$ by u_ν and taking the logarithm of this quantity, we have

---

[1] See Picard: Cours d'Analyse, T. II., p. 145–149. The function considered by Picard differs from the above in having the essentially singular points condensed on the unit circle so that the circle becomes an essentially singular line of the first category and not of the second category. This difference is all important, since clearly Picard's function can present no such property as the one with which we are chiefly concerned in this paper.

$$\log u_\nu = \log \frac{z-(1+a_\nu)e^{\nu i}}{z-(1+b_\nu)e^{\nu i}} + \frac{(a_\nu-b_\nu)e^{\nu i}}{z-(1+b_\nu)e^{\nu i}} + \frac{1}{2}\left[\frac{(a_\nu-b_\nu)e^{\nu i}}{z-(1+b_\nu)e^{\nu i}}\right]^2$$
$$+ \dots + \frac{1}{\nu-1}\left[\frac{(a_\nu-b_\nu)e^{\nu i}}{z-(1+b_\nu)e^{\nu i}}\right]^{\nu-1},$$

where $\log \frac{z-(1+a_\nu)e^{\nu i}}{z-(1+b_\nu)e^{\nu i}}$ is a uniform function within the unit circle. Developing we find

$$\log u^\nu = -\frac{(a_\nu-b_\nu)e^{\nu i}}{z-(1+b_\nu)e^{\nu i}} + \frac{1}{2}\left[\frac{(a_\nu-b_\nu)e^{\nu i}}{z-(1+b_\nu)e^{\nu i}}\right]^2$$
$$-\dots-\frac{1}{\nu-1}\left[\frac{(a_\nu-b_\nu)e^{\nu i}}{z-(1+b_\nu)e^{\nu i}}\right]^{\nu-1} - \frac{1}{\nu}\left[\frac{(a_\nu-b_\nu)e^{\nu i}}{z-(1+b_\nu)e^{\nu i}}\right]^{\nu}$$
$$+\frac{(a_\nu-b_\nu)e^{\nu i}}{z-(1+b_\nu)e^{\nu i}} + \dots .$$

We have, therefore,

$$\log \psi(z) = -\sum \frac{1}{\nu}\left[\frac{(a^\nu-b^\nu)e^{\nu i}}{z-(1+b_\nu)e^{\nu i}}\right]^{\nu} +$$
$$\frac{1}{\nu+1}\left[\frac{(a_\nu-b_\nu)e^{\nu i}}{z-(1+b_\nu)e^{\nu i}}\right]^{\nu+1} + \dots,$$

from which we derive

$$e^{\log\psi(z)} = \psi(z) = e^{-\sum \frac{1}{\nu}\left[\frac{(a_\nu-b_\nu)e^{\nu i}}{z-(1+b_\nu)e^{\nu i}}\right]^{\nu} +}$$
$$\frac{1}{\nu+1}\left[\frac{(a_\nu-b_\nu)e^{\nu i}}{z-(1+b_\nu)e^{\nu i}}\right]^{\nu+1} + \dots,$$

and we have to prove now that the series

$$(11) \quad -\sum\left[\frac{1}{\nu}\frac{(a_\nu-b_\nu)e^{\nu i}}{z-(1+b_\nu)e^{\nu i}}\right]^{\nu} +$$
$$\frac{1}{\nu+1}\left[\frac{(a_\nu-b_\nu)e^{\nu i}}{z-(1+b_\nu)e^{\nu i}}\right]^{\nu+1} + \dots$$

is finite and continuous in the part of the plane where $|z-(1+b_\nu)e^{\nu i}| > |a_\nu - b_\nu|$ say for instance the unit circle around the origin. The general term of (11) is

$$f_\nu(z) = \frac{1}{\nu}\left[\frac{(a_\nu-b_\nu)e^{\nu i}}{z-(1+b_\nu)e^{\nu i}}\right]^{\nu} + \frac{1}{\nu+1}\left[\frac{(a_\nu-b_\nu)e^{\nu i}}{z-(1+b_\nu)e^{\nu i}}\right]^{\nu+1} + \dots .$$

Replacing each term by its modulus and z by $|z|$, we get

$$(12)\quad |f_\nu(|z|)| < \frac{1}{\nu}\left[\frac{|a_\nu - b_\nu|}{1 + b_\nu - |z|}\right]^\nu + \frac{1}{\nu+1}\left[\frac{|a_\nu - b_\nu|}{1 + b_\nu - |z|}\right]^{\nu+1} + \dots$$

$$< \left[\frac{|a_\nu - b_\nu|}{1 + b_\nu - |z|}\right]^\nu \frac{1}{1 - \dfrac{|a_\nu - b_\nu|}{1 + b_\nu - |z|}}.$$

But the series whose general term is equal to the right-hand side of (12) is absolutely convergent; in fact, we have

$$\lim_{\nu=\infty} \sqrt[\nu]{\left\{\frac{|a_\nu - b_\nu|}{1 + b_\nu - |z|}\right\}^\nu} = \lim_{\nu=\infty} \frac{|a_\nu - b_\nu|}{1 + b_\nu - |z|} < 1$$

for values of z such that $|z - (1 + b_\nu)e^{\nu i}| < |a_\nu - b_\nu|$. We shall now proceed to investigate the behavior of $\psi(z)$ on the unit circle and we propose to find, just as before, the necessary and sufficient condition which the ensembles a_ν and b_ν must fulfil in order that $\psi(z)$ and all its derivatives shall be finite and continuous within and on the unit circle.

Putting $|z| = 1$ in (12) we have

$$(13)\qquad |f_\nu(z)| \underset{|z|=1}{<} \frac{1}{\nu}\left[\frac{a_\nu - b_\nu}{b_\nu}\right]^\nu \frac{1}{1 - \dfrac{a_\nu - b_\nu}{b_\nu}}$$

and the series whose general term is equal to the right-hand side of (13) will obviously be convergent if

$$\frac{a_\nu - b_\nu}{b_\nu} < 1$$

which therefore is the condition that must be fulfilled in order that $\psi(z)$ shall be finite for $|z| = 1$. Let us now write

$$\psi(z) = e^{\phi(z)},$$

where $\phi(z)$ is identical with the series (11). We find for the successive derivatives of $\psi(z)$

$$\psi'(z) = \psi(z)\phi'(z)$$
$$\psi''(z) = \psi'(z)\phi(z) + \psi(z)\phi''(z)$$
$$\cdots\cdots$$
$$\cdots\cdots$$
$$\psi^n(z) = \psi^{n-1}(z)\phi'(z) + (n-1)\psi^{n-2}(z)\phi''(z) + \ldots + \psi(z)\phi^n(z).$$

These expressions show that, if $\psi(z)$ be finite on the unit circle, the successive derivatives of $\psi(z)$ will have the same property provided also we can make the successive derivatives of $\phi(z)$ finite on this line. Hence to find the condition which must be satisfied in order that $\psi(z)$ as well as all its derivatives up to any order as high as we please shall be finite and continuous inwardly on the unit circle is manifestly equivalent to finding the condition that $\phi(z)$ shall possess the same property.

We now have

$$\phi'(z) = \sum \frac{1}{z-(1+b_\nu)e^{\nu i}} \left[\frac{(a_\nu - b_\nu)e^{\nu i}}{z-(1+b^\nu)e^{\nu i}}\right]^\nu \times$$
$$\left[1 + \frac{(a_\nu - b_\nu)e^{\nu i}}{z-(1+b_\nu)^{\nu i}} + \ldots\right]$$

which series, since the part in bracket is itself a convergent series, evidently can be put into the form

$$\phi'(z) = \sum \left\{\frac{(a_\nu - b_\nu)e^{\nu i}}{z-(1+b_\nu)e^{\nu i}}\right\}^\nu \frac{1}{z-(1+b_\nu)e^{\nu i}} \frac{1}{1-\dfrac{(a_\nu - b_\nu)e^{\nu i}}{z-(1+b_\nu)e^{\nu i}}}.$$

In the same way we find

$$\phi''(z) = -\sum \left\{\frac{(a_\nu - b^\nu)e^{\nu i}}{z-(1+b_\nu)e^{\nu i}}\right\}^\nu \left[\frac{1}{z-(1+a_\nu)e^{\nu i}}\right] \times$$
$$\left[\frac{\nu}{z-(1+b_\nu)e^{\nu i}} + \frac{1}{z-(1+a_\nu)e^{\nu i}}\right]$$

and in general

$$\phi^n(z) = (-1)^{n-1} \sum \left\{\frac{a_\nu - b_\nu)e^{\nu i}}{z-(+b_\nu)e^{\nu i}}\right\}^\nu \frac{1}{z-(1+a_\nu)e^{\nu i}} \times$$
$$\left[\frac{\nu.\nu+1 \ldots \nu+n-2}{[z-(1+b_\nu)e^{\nu i}]^{n-1}} + \ldots +\right.$$

$$(n-1)\frac{\nu.\,\nu+1\ldots\nu+n-3}{[z-(1+b_\nu))e^{\nu i}]^{n-2}[z-(1+a_\nu)e^{\nu i}}+\cdots$$

$$+\frac{\underline{n-1|}}{[z-(1+a_\nu)c^{\nu i}]^{n+1}}\Big].$$

Replacing in this series each term by its modulus and putting $z=|z|=1$ we have

$$|\phi^n(z)|_{|z|=1}<\sum\frac{|a_\nu-b_\nu|^\nu}{b_\nu^\nu}\cdot\frac{1}{a_\nu}\Big[\frac{\nu.\,\nu+1\ldots\nu+n-2}{b_\nu^{n-1}}+$$

$$\cdots+(n-1)\frac{\nu.\,\nu+1\ldots\nu+n-3}{b_\nu^{n+2}a_\nu}+\cdots+\frac{\underline{n-1|}}{a_\nu^{n-1}}\Big].$$

If in this series we replace each numerical coefficient of the terms in bracket by the greatest of them, *viz.*, $\nu.\,\nu+1\ \ldots\ \nu+n-2$, we get

$$|\phi^n(z)|_{|z|=1}<\sum\left|\frac{a_\nu-b_\nu}{b_\nu}\right|\frac{1}{a_\nu}\Big[\frac{1}{b_\nu^{n-1}}+\frac{1}{b_\nu^{n-2}a_\nu}+\cdots$$

$$\cdots+\frac{1}{a_\nu^{n-1}}\Big](\nu.\,\nu+1\ldots\nu+n-2)$$

$$<\underline{n-1|}\sum\left|\frac{a_\nu-b_\nu}{b_\nu}\right|^{\nu-1}\Big[\left(\frac{1}{b^\nu}\right)^n-\left(\frac{1}{a_\nu}\right)^n\Big]\underline{\nu|},$$

since

$$\frac{\underline{\nu+n-2|}}{\underline{\nu-1|}}<\underline{\nu|}\ \ \underline{n-1|}.$$

In order that this last series shall be absolutely convergent, we must have

$$\lim.\left|\frac{a_\nu-b_\nu}{b_\nu}\right|\sqrt[\nu]{\left(\frac{1}{b_\nu}\right)^n-\left(\frac{1}{a_\nu}\right)^n}\cdot\sqrt[\nu]{\underline{\nu|}}<1;$$

but we have

$$\sqrt[\nu]{\left|\left(\frac{1}{b_\nu}\right)^n-\left(\frac{1}{a_\nu}\right)^n\right|}<\sqrt[\nu]{\left(\frac{1}{a_\nu}\right)^n}$$

and therefore

$$\lim. \left|\frac{a_\nu - b_\nu}{b_\nu}\right| \sqrt[\nu]{\left|\left(\frac{1}{b_\nu}\right)^n - \left(\frac{1}{a_\nu}\right)^n\right|}$$

$$< \lim. \left|\frac{a_\nu - b_\nu}{b_\nu}\right|^{\nu-1} \sqrt[\nu]{\left(\frac{1}{a_\nu}\right)^n};$$

we have also

$$\lim. \sqrt[\nu]{\nu!} = \lim_{\nu=\infty} \sqrt[\nu]{\sqrt{2\pi}\, e^{-\nu + \frac{1}{12\nu}} \cdot \nu^{\nu + \frac{1}{\nu}}} = \lim_{\nu=\infty} \frac{\nu}{e}$$ [1]

and we find the condition

$$\lim_{\nu=\infty} \left|\frac{a_\nu - b_\nu}{b_\nu}\right| \sqrt[\nu]{\left(\frac{1}{a_\nu}\right)^n} \sqrt[\nu]{\nu!} = \lim. \left|\frac{a_\nu - b_\nu}{b_\nu}\right| \sqrt[\nu]{\left(\frac{1}{a_\nu}\right)^n} \cdot \frac{\nu}{e} < 1,$$

which will be satisfied for all values of n if we put $n = \nu$. We have therefore finally the condition

$$(14) \qquad \frac{a_\nu - b_\nu}{a_\nu b_\nu} < \frac{e}{\nu}.$$

If then the ensembles a_ν and b_ν be chosen in such a way as to satisfy the condition (14), all the successive derivatives of $\phi(z)$ up to any order as high as we please will be finite on the unit circle and the relations on page 43 show that $\psi(z)$ and all its derivatives will possess the same property. Examples of such ensembles are

$$(a) \quad a_\nu = \frac{1}{\nu},\ b_\nu = \frac{1}{\nu - \frac{1}{\nu}};\ a_\nu = \frac{1}{2^\nu},\ b_\nu = \frac{1}{2^\nu + \frac{2}{\nu}},\ \text{etc.}$$

$$(b) \quad a_\nu = \frac{1}{\nu},\ b_\nu = \frac{1}{\nu} - \frac{1}{\nu^3};\ a_\nu = \frac{1}{2^\nu},\ b_\nu = \frac{1}{2^\nu} + \frac{1}{2^{2\nu}},\ \text{etc.}$$

If we write the ensembles a_ν and b_ν in the form

$$a_\nu = \frac{1}{a_\nu'},\quad b_\nu = \frac{1}{a_\nu' \pm b_\nu'},$$

the condition (14) reduces to

$$b_\nu' < \frac{e}{\nu}.$$

[1] See Serret : Cours de Calcul diff. and integ., T. II, p. 208.

which is seen to be fulfilled by the ensembles (a) given above. A second form of ensembles is

$$a_\nu = \frac{1}{a_\nu'},\ b_\nu = \frac{1}{a_\nu'} \pm \frac{1}{b_\nu'},$$

which gives us the condition

$$\left| a_\nu' - a_\nu' \left\{ \frac{1}{1 \pm \frac{a_\nu'}{b_\nu'}} \right\} \right| = \left| a_\nu' - a_\nu'\left(1 \pm \frac{a_\nu'}{b_\nu'} \mp \cdots\right) \right| < \frac{a_\nu'^2}{b_\nu'} < \frac{e}{\nu},$$

which is seen to be satisfied by the ensembles (b) written above.

We may state the result of our investigation thus:

It is always possible to choose ensembles a_ν *and* b_ν *such that the function*

$$\psi(z) = \prod \frac{z - (1 + a_\nu)e^{\nu i}}{z - (1 + b_\nu)e^{\nu i}} e^{\Phi_\nu(z)}$$

as well as all its derivatives up to any order as high as we please shall be finite and inwardly continuous within and on the unit circle admitting this as a doubly singular line.

We found above the following inequality:

$$|\Phi^n(z)| < \underline{n-1|} \sum \left| \frac{a_\nu - b_\nu}{b_\nu} \right|^{\nu - 1} \left| \left(\frac{1}{b_\nu}\right)^n - \left(\frac{1}{a_\nu}\right)^n \underline{\nu|} \right|,$$

and we notice that the presence of the factors $\underline{\nu|}$ and $\left|\frac{a_\nu - b_\nu}{b_\nu}\right|^{\nu-1}$ is due entirely to the fact that $\psi(z)$ has the exponential factor $e^{\Phi_\nu(z)}$ affixed. If this factor is removed, we get the function F(z) considered above; but this amounts simply to putting $\nu = 0$ in these factors, since $\underline{0|} = 1$ and $\left|\frac{a_\nu - b_\nu}{b_\nu}\right|^0 = 1$.

Doing this we get the inequality

$$|\phi^n(z)| < \underline{n-1|} \sum \left| \left(\frac{1}{b_\nu}\right)^n - \left(\frac{1}{a_\nu}\right)^n \right| = \text{an abs. conv.},$$

a series which is nothing but the condition obtained on p. 37 for F(z).

If we study the ensembles given on page 39, we notice that the poles and zeros approach each other more rapidly as we get nearer and nearer the unit circle. On the other hand, the series of ensembles on page 45 ((*a*) and (*b*)) present a much slower approach of zeros and essentially singular points. It is not without interest to notice the following fact :

In order that the functions $F(z)$ and $\psi(z)$ and all their derivatives shall be finite and continuous on the unit circle, it is necessary and sufficient that their respective logarithmic derivatives shall possess the same property.

Suppose now we write $F(z)$ in the form

$$F(z)=\prod_{1}^{\nu}(1-f_\nu(z))=\prod_{1}^{\nu}\left(1-\frac{(a_\nu-b_\nu)e^{\nu i}}{z-(1+b_\nu)e^{\nu i}}\right).$$

We may easily prove that if $\Sigma f_\nu(z)$ and all its derivatives are finite and continuous on the unit circle, the same will hold for $F(z)$ and all its derivatives. In fact, the necessary and sufficient condition for this is

$$\left|\underline{n|}(-1)^n\sum\frac{(a_\nu-b_\nu)e^{\nu i}}{\{z-(1+b_\nu)e^{\nu i}\}^{n+1}}\right|<\underline{n|}\sum\frac{|a_\nu-b_\nu|}{\{|z|-(1+b_\nu)\}^{n+1}}$$

$$<\underline{n|}\sum\left|\frac{a_\nu-b_\nu}{b_\nu^{n+1}}\right|=\text{an abs. conv. series.}$$

for all values of n however large. But, since $\lim\limits_{\nu=\infty}\frac{a_\nu}{b_\nu}=1$, the above series will be convergent whenever the series

$$\sum\left[\left(\frac{1}{b_\nu}\right)^n-\left(\frac{1}{a_\nu}\right)^n\right]$$

is convergent and conversely. Hence, since this is nothing but the condition that $F(z)$ shall be finite on the unit circle as well as all its derivatives, our proposition is established.

We are, by means of this proposition, enabled to establish a one-to-one correspondence between the types of functions discussed by Pringsheim and functions like $F(z)$. In fact, given a function

$$f(z)=\sum\frac{c_\nu}{z-(1+b_\nu)e^{\nu i}}$$

where $\sum \frac{c_\nu}{b_\nu{}^{n+1}}$ is an absolutely convergent series for all values of n, if we write

$$a_\nu = b_\nu + c_\nu$$

the function

$$F(z) = \prod \left(1 - \frac{(a_\nu - b_\nu)e^{\nu i}}{z - (1 + b_\nu)e^{\nu i}}\right).$$

which has the same poles as $f(z)$ and whose zeros are the points of the ensemble $a_\nu' = (1 + b_\nu + c_\nu)e^{\nu i}$ will, by the above proposition possess the same property as $f(z)$ with respect to the unit circle. Thus suppose we start with the function

$$f(z) = \sum \frac{1}{\nu^{2\nu}} \frac{1}{z - \left(1 + \frac{1}{\nu}\right) e^{\nu i}};$$

we put $a_\nu = 1 + \frac{1}{\nu} + \frac{1}{\nu^{2\nu}}$ making the series $\sum \frac{\nu^{n+1}}{\nu^{2\nu}}$ absolutely convergent for all values of n and the corresponding function $F(z)$ will be

$$F(z) = \prod_1^\nu \frac{z - \left(1 + \frac{1}{\nu} + \frac{1}{\nu^{2\nu}}\right) e^{\nu i}}{z - \left(1 + \frac{1}{\nu}\right) e^{\nu i}}.$$

VITA.

The author, John Eiesland, was born in 1867 near Kristianssand, Norway. He received his elementary instruction in the public school. After having completed in 1884 a course at a normal school, he attended for some time Kristianssand's Latin and Real Skole. He afterwards engaged in teaching up till 1888, when he came to the United States. In January, 1889, he entered the State University of South Dakota, from which institution he graduated in 1891 with the degree of Ph.B. From 1891–1892 he taught mathematics and sciences in a private academy in Minnesota. In the Fall of 1892 he entered the Johns Hopkins University, where for three years he pursued the study of mathematics with physics and astronomy as minor subjects.

In 1895 he was appointed professor of mathematics in Thiel College, Pa., which position he still holds, having a leave of absence while completing his course. During the past year he has held a university scholarship in mathematics.

In conclusion he wishes to express his gratitude to Professors Craig and Hulburt for the interest taken in his work while a student at the university.

www.ingramcontent.com/pod-product-compliance
Lightning Source LLC
Chambersburg PA
CBHW032117080426
42733CB00008B/969
9783337179021